j594
.56
C5790

W9-DDA-215

WILDER BRANCH LIBRARY
7140 E. SEVEN MILE RD.
DETROIT, MI 48234

9/04
WI

Octopuses

Octopuses

Anna Claybourne

Chicago, Illinois

© 2004 Raintree
Published by Raintree, a division of Reed Elsevier, Inc.
Chicago, Illinois
Customer Service 888-363-4266
Visit our website at www.raintreelibrary.com

All rights reserved. No part of this book may be reproduced or utilized in any form or by any means, electronic or mechanical, including photocopying, recording, taping, or any information storage and retrieval system, without permission in writing from the publisher. For information, address the publisher:
Raintree, 100 N. LaSalle, Suite 1200, Chicago, IL 60602

Project Editors: Geoff Barker, Marta Segal Block, Jennifer Mattson, Tamsin Osler
Production Manager: Brian Suderski
Illustrated by Peter Bull Art Studio
Map art by Stefan Chabluk
Designed by Ian Winton
Consultants: Michael Chinery and William Wardle
Picture research by Vashti Gwynn
Planned and produced by Discovery Books
Printed and bound in the United States by Lake Book Manufacturing, Inc.
07 06 05 04 03
10 9 8 7 6 5 4 3 2 1

Library of Congress Cataloging-in-Publication Data:
Claybourne, Anna.
Octopuses / Anna Claybourne.
v. cm. -- (The secret world of)
Includes bibliographical references and index.
Contents: Life with eight legs -- Where octopuses live -- Octopus senses -- Finding food -- Staying safe -- Reproduction -- Octopus intelligence -- Octopuses and people.
ISBN 0-7398-7024-6 (lib. bdg.-hardcover)
1. Octopodidae--Juvenile literature. [1. Octopus.] I. Title. II. Series.
QL430.3.O2C63 2003
594'.56--dc21
2003001916

Acknowledgments
The publishers would like to thank the following for permission to reproduce photographs:
p.9 Reinhard Dirsheri/Ecoscene; p.10 Pacific Stock/Bruce Coleman Collection; p.11A B. Jones & M. Shimlock/Natural History Photographic Agency; p.11B Manuel Bellver/Corbis; pp. 12, 24B, 32, 33 Norbert Wu/Natural History Photographic Agency; p.14 Laurence Gould/Oxford Scientific Films; p.15 Karen Gowlett-Holmes/Oxford Scientific Films; p.16 Kjell Sandved/Ecoscene; pp.18, 34 Jane Burton/Bruce Coleman Collection; pp.19, 26, 35 Jeffrey L. Rotman/Corbis; p.20 Mark Deeble & Victoria Stone/Oxford Scientific Films; p.23 Daniel Heuclin/Natural History Photographic Agency; p.24A Trevor McDonald/Natural History Photographic Agency; p.25 Oxford Scientific Films; p.27 Kim Westerkov/Oxford Scientific Films; p.28 TammyPeluso/Oxford Scientific Films; p.29 D. Fleetham/Silvestris/Frank Lane Picture Agency; p.30 Tony Ayling/Natural Visions; pp.31, 41B Frank Lane Picture Agency; p.36 Silvestris Fotoservice/Frank Lane Picture Agency; p.37 Mauro Fermariello/Science Photo Library; p.38 K. Aitken/Frank Lane Picture Agency; p.39 Anne Norris/Natural Visions; p.40 Lawrence Gresswell/Eye Ubiquitous/Corbis; p.41A Luc Cuyvers/Corbis; p.42 Rudie Kuiter/Oxford Scientific Films; p.43 Rudie Kuiter/Oxford Scientific Films.

Other Acknowledgments
Front cover: Oxford Scientific Films

Disclaimer
Every effort has been made to contact copyright holders of any material reproduced in this book. Any omissions will be rectified in subsequent printings if notice is given to the publisher.

Note to the Reader
Some words are shown in bold, **like this.** You can find out what they mean by looking in the glossary.

Contents

CHAPTER 1
Life with Eight Legs

There are about 300 different kinds of octopus. Scientists are still discovering new ones.

The biggest octopus is the giant octopus. From the tip of one leg to the tip of the opposite leg, an average giant octopus measures about 10 ft (3 m). Some grow to more then 23 ft (7 m) across.

Several types of octopus weigh less than 0.04 oz (1 g). An octopus this small could easily sit on your fingertip.

The word *octopus* comes from Greek and Latin, and means "eight-footed." Although some people use the word *octopi* when talking about more than one octopus, most scientists today prefer to say *octopuses*.

The octopus is an unusual-looking sea creature. Its body is made up of a head with two eyes, eight legs, and a squishy, bag-shaped structure called the **visceral mass.** Most animals have their heads at one end of their bodies, but an octopus's head is right in the middle. It is between its legs and the visceral mass.

Octopuses are **invertebrates,** which means that the body of an octopus has no backbone. In fact, octopuses have no bones at all. The only hard part of an octopus is its parrotlike **beak,** located in the middle of its legs. Octopuses are related to other soft, squishy animals such as slugs. Although they have no skeleton to keep their bodies from collapsing, octopuses can grow very large because they live in the ocean. The water supports their soft bodies.

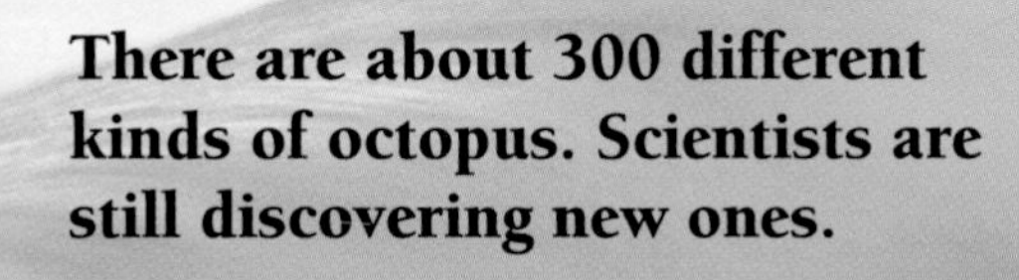

Suckers
These are used to grip smooth surfaces and to hold objects firmly.

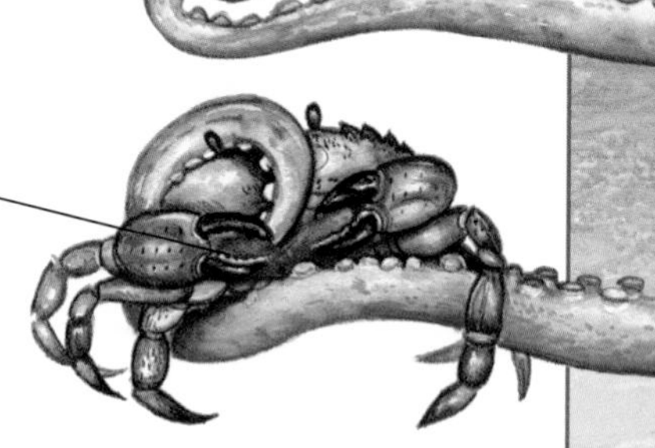

Legs
Octopuses use their legs for moving around and for catching prey.

Octopuses are bigger and much more intelligent than most other invertebrates. They can use their legs to pick up objects, and they are very good at learning new things. They are also able to change the color of their skin like chameleons.

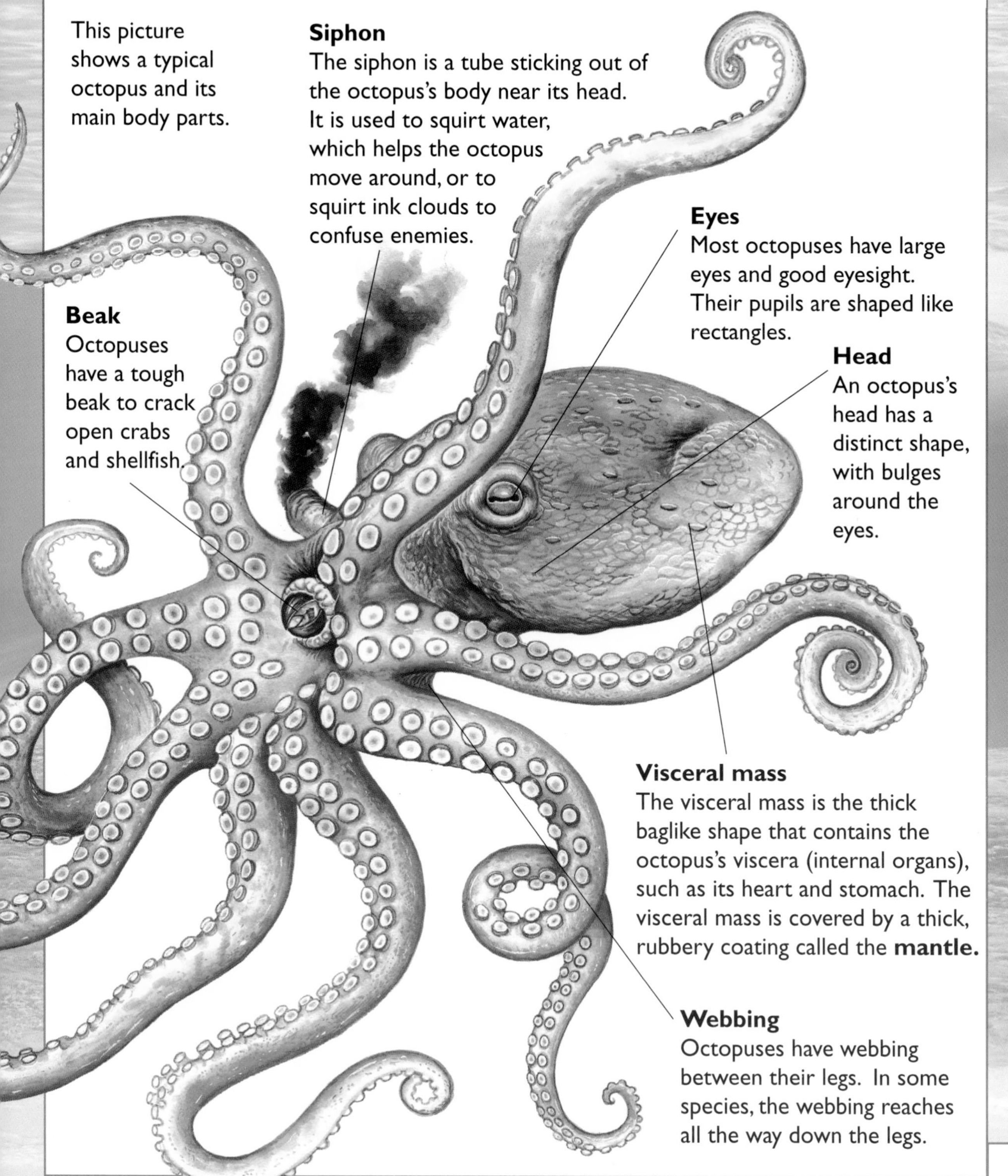

Inside an Octopus

Although octopuses do not have any bones, they have a complex system of hundreds of crisscrossing muscles. These allow the octopus to control its eight legs separately and to make precise movements. Octopuses are very strong. In tests, some octopuses moved objects that weighed up to 50 times more than themselves. Unlike most other animals, they use muscles instead of fat to store energy. Octopuses have hardly any fat in their bodies.

Most of an octopus's internal organs are inside its **visceral mass.** It has some of the same organs as humans do, such as digestive organs and a liver. Instead of lungs, though, octopuses have organs called **gills.** An octopus breathes by sucking in water through its **siphon.** The water fills up a chamber in part of its visceral mass called the **mantle cavity.** Then it flows past the gills, which soak up oxygen from the water into the octopus's blood. The octopus squirts the used water out through its siphon.

Another organ inside the octopus's visceral mass is the **ink sac.** It produces a black, inky liquid, which the octopus squirts out through its siphon to startle and confuse enemies.

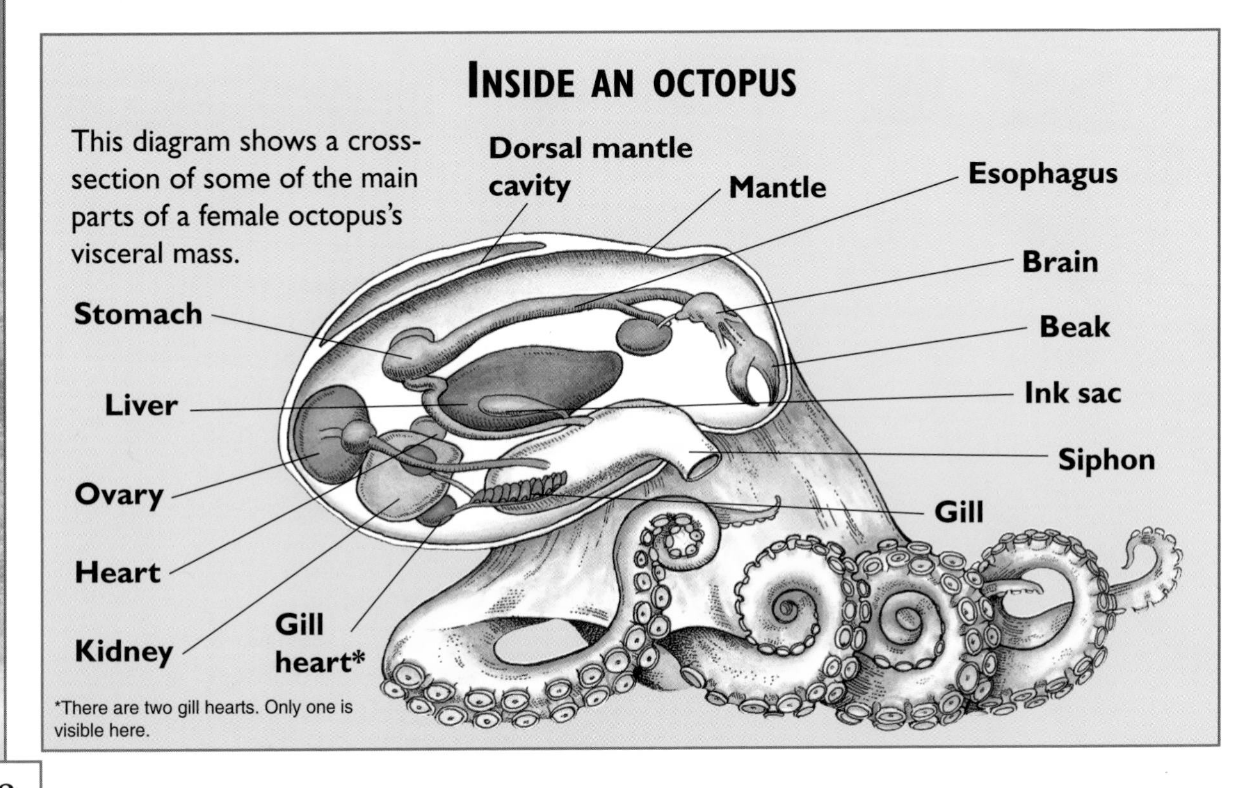

Jet Propelled

Octopuses normally move around slowly by crawling along the ocean floor. But if an octopus needs to make a quick getaway, it can shoot water out of its siphon for a burst of jet-propelled power. In the picture below, a common octopus zooms along using this method.

Octopuses have three hearts. The main heart pumps blood around the body, while the two smaller hearts pump blood past the gills. Octopus blood is not red, but bright blue. This is because unlike our blood, which contains iron, octopus blood contains a lot of copper. Like all **invertebrates,** octopuses are **ectothermic,** also known as cold-blooded. This means that they cannot warm their own bodies up, but stay more or less at the same temperature as their surroundings.

The octopus's brain is in its head. In proportion to its body size, an octopus's brain is much bigger than most other invertebrates' brains.

Octopus Relatives

Octopuses belong to a class of sea creatures called **cephalopods.** The word cephalopod means "head-foot" in Greek. This name makes sense, because cephalopods have their heads attached directly to their feet. Other cephalopods include squid, cuttlefish, and nautiluses.

All cephalopods have many legs and a bag-shaped or torpedo-shaped **visceral mass** covered by a **mantle**. They have good eyesight, large brains, and unusually high levels of intelligence. Cephalopods belong to a larger animal group called **mollusks**, which also includes clams, oysters, and some land animals such as slugs and snails. However, cephalopods are different from the majority of other mollusks because most cephalopods do not have a shell.

Squid and cuttlefish are very similar to each other. They both have long, torpedo-shaped bodies and ten limbs—eight legs with **suckers,** and two longer **tentacles** with paddle-shaped ends, which they use for catching food.

This is a reef squid from Indonesia in Southeast Asia. A squid's eyes are very big in comparison to the rest of its body.

This is a type of nautilus that lives in the southern Pacific Ocean.

Squid and cuttlefish do not crawl around as much as octopuses do. Instead, they use their **siphons** to move very quickly as they chase fish and other **prey.** Giant squid can grow to a length of more than 50 feet (15 meters). Their eyeballs, up to 11 inches (28 centimeters) across, are the biggest of any animal in the world. Like octopuses, they are good at changing color.

Nautiluses do not look like the other cephalopods. A nautilus has a coiled shell like a snail's shell, and as many as 94 legs.

Cephalopods Ruled the Seas

I DIDN'T KNOW THAT!

Four hundred million years ago, cephalopods were the dominant life form in the ocean. There were more cephalopods than fish. Many of them were very large, and as hunters, they played a role in the ancient oceans similar to that of today's large ocean predators, such as sharks.

The photograph below shows ammonites, the fossilized shells of a type of prehistoric cephalopod. Millions of years ago, many kinds of cephalopods had shells, but today only nautiluses still have them.

CHAPTER 2
Where Octopuses Live

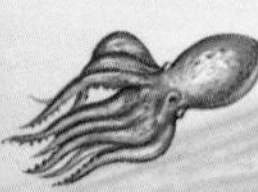

Most octopuses live to be only one or two years old.

Some octopuses, such as the common octopus, move around between different depths of water depending on the time of year. They live in deeper water in the winter and move back to shallow water to breed.

Octopuses are found in seas and oceans all over the world. Most **species** are **benthic,** which means that they spend a lot of time on the ocean floor. They tend to prefer warm, shallow water near the shore. The very poisonous blue-ringed octopus lives in rock pools and shallow waters off the coast of Australia, while the Caribbean reef octopus lives among coral reefs in the warm, tropical waters of the Caribbean.

Reef Dwellers

Many species of octopus like living in coral reefs. These are large, rocky underwater structures built over

An Antarctic octopus in an underwater ice cave searches for fish and crabs to eat.

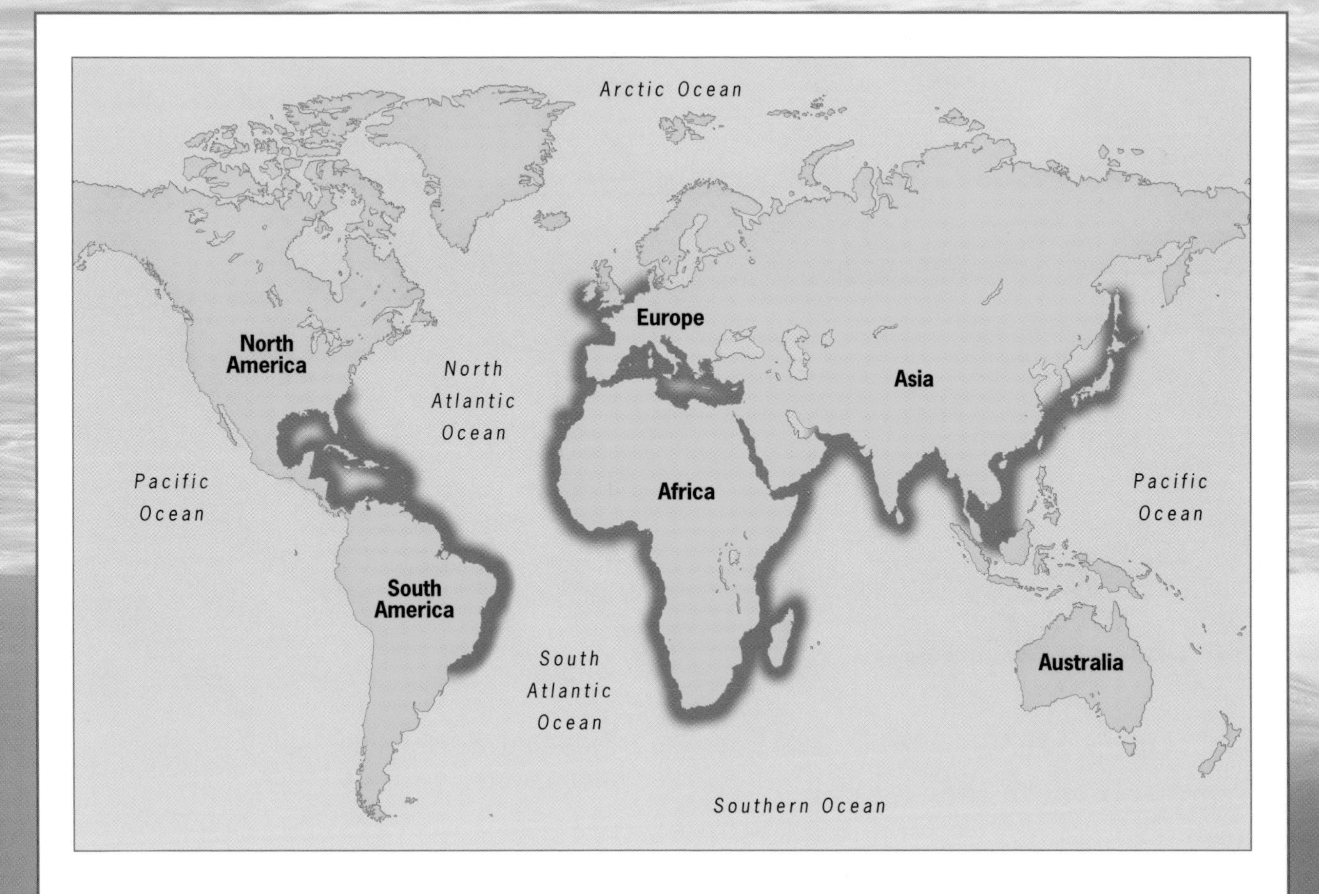

This map shows where the common octopus is found. It lives mostly in coastal areas and avoids cold places.

hundreds of years by millions of coral polyps, tiny sea creatures that are often just called coral. Coral reefs have many caves and hollows that make good hiding places for octopuses. Reefs also provide homes for clams, crabs, and sea snails, which octopuses like to eat.

Deep-Sea Homes

There are also some deep-sea octopus species and some that live around the poles. The deep-sea North Atlantic octopus, which is about the size of a human hand, lives at depths of up to nearly 2,000 feet (600 meters) in the North Atlantic Ocean. The giant octopus can be found as far north as Alaska. A tiny octopus called the Antarctic octopus lives in the icy waters of the Southern Ocean.

Octopuses can only live in salt water, so they are never found in lakes, ponds, or rivers. Octopuses that are kept in aquariums or science labs have to live in water that is kept at the right temperature and level of salinity, or saltiness.

A common octopus peeks out from its den under a seaweed-covered rock.

Octopus Dens

Perhaps because they do not have a protective shell of their own, octopuses like to have a den to live in. They feel safest in a tight, enclosed space that is just big enough for them to fit inside.

An octopus's den can be a crack in the ocean floor, a cave or hollow in a coral reef, or a space between rocks. Octopuses sometimes dig themselves a hole in the ocean floor under a rock to make a den. They may also pile up rocks around the outside, along with broken bits of shells from the animals they have eaten. These piles of leftovers are called **middens** and they are very useful to scientists. Octopuses are very good at hiding, so they are almost impossible to spot in the wild. If scientists see a midden, they know that an octopus den is probably not far away.

Home Alone

Although some **species** of octopus have been known to get together in groups, they usually live alone. They spend most of their time inside their dens, going out occasionally to hunt for food. They often bring food back to their den to eat it. Octopuses usually position themselves in their dens so that one eye is peeping out to watch for danger or **prey.**

Secondhand Homes

I DIDN'T KNOW THAT!

Smaller octopuses, such as pygmy octopuses, often live inside shells abandoned by other creatures. Sometimes they move into an old oil drum, a glass jar, or even a broken beverage can, as in this picture of a sand octopus from Australia. Octopuses do not mind what their den is made of, as long as it protects them and makes them feel safe.

Unfortunately for octopuses, their fondness for containers makes it easy for humans to trick them. People who fish for octopuses place special pots on the ocean floor. Octopuses often use the pots as dens, and then the humans simply come back to collect the pots with the octopuses inside.

CHAPTER 3
Octopus Senses

Like other **cephalopods,** nearly all octopuses have very good eyesight. In fact, sight is an octopus's most important sense.

While most **invertebrates** have very simple eyes, octopus eyes are more like our own. Octopuses can reduce and enlarge their pupils (the openings in the middle of the eyes) to let in different amounts of light, and they can **focus** on different objects. However, while humans focus by changing the thickness of the lens at the front of the eye, an octopus focuses by moving its lens backward and forward. An octopus's **retina,** the patch at the back of the eye that

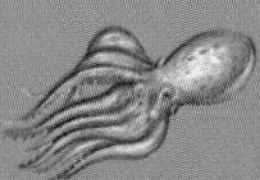

Scientists have found that octopuses have a favorite eye. They are either left-eyed or right-eyed, just as humans can be left-handed or right-handed.

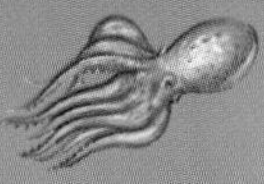

The biggest octopus eyes are the size of car headlights.

Octopuses have a good sense of touch, but they cannot tell the difference between heavy and light objects.

A giant octopus may have more than 2,000 suckers covering its legs.

A close-up photo of the eye of the Caribbean reef octopus.

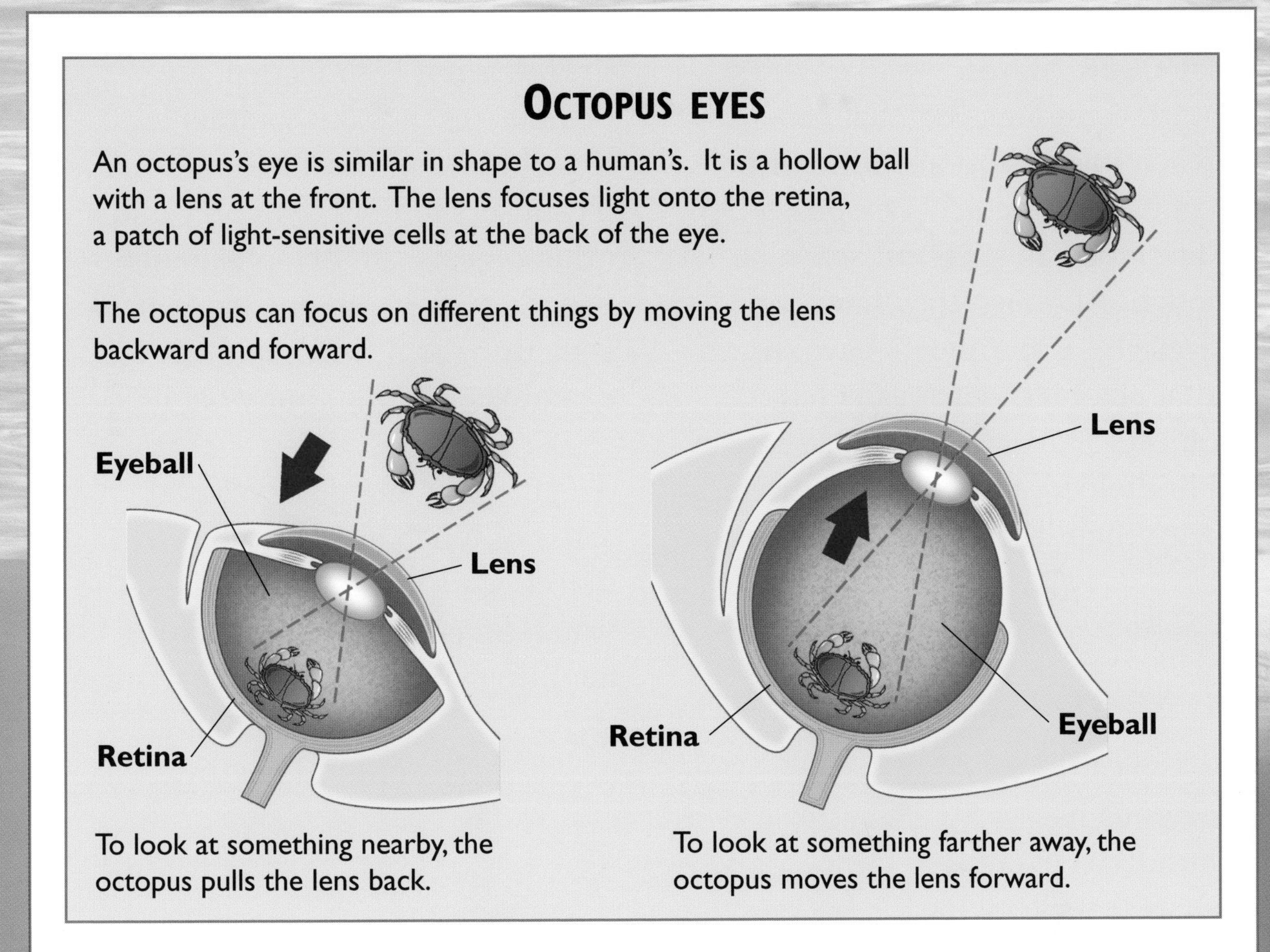

To look at something nearby, the octopus pulls the lens back.

To look at something farther away, the octopus moves the lens forward.

detects light, is extremely sensitive. Each retina contains 20 million light-sensitive cells.

Octopuses can detect polarized light, which is light that has been slightly changed after passing through a transparent object. Polarized light looks just like normal light to humans, but to octopuses, it looks different. The ability to see polarized light helps octopuses to spot transparent animals moving around in the water, such as jellyfish.

The Blind Octopus

The blind octopus is named for its very poor eyesight. It lives so deep in the Atlantic Ocean that there is no sunlight. The eyes of the blind octopus have no lenses, and the retina hardly works at all.

However, scientists think that the octopus is not completely blind, because it seems to be able to tell the difference between light and dark. It belongs to a group of octopuses that have large fins to help them swim.

Sucker Sense

With eight long legs that can reach into almost any kind of hole, container, or crack, octopuses have developed an excellent sense of touch. Although they have good eyesight, experiments have shown that octopuses can easily tell objects apart when they have been blindfolded, just by feeling them with their legs and their **suckers.**

Even when it is in its den, an octopus reaches out with its legs to feel for food nearby. It also uses its suckers to hold onto prey and other objects. An octopus might use the suckers on one leg to hold something still, while another leg examines it to find out what it is.

An octopus's highly developed sense of touch also allows it to detect ripples and vibrations in the water. So even without touching anything, an octopus can feel other animals moving around nearby.

A typical octopus has hundreds of cup-shaped suckers. Each sucker has thousands of sensitive nerve endings, especially around the outside edge. The size of an octopus's suckers depends on the size of its overall body. The giant octopus has the biggest suckers. They can be as large as a small plate.

Most octopuses, like the white-spotted octopus shown here, have two rows of suckers on each leg.

A giant octopus clings on firmly to a cabezon fish that it has captured.

How Suckers Suck

An octopus holds things with its suckers by clamping them onto an object, then pulling upward on the middle of the sucker to create a **vacuum** underneath it. This reduces the pressure in the space under the sucker. The greater pressure outside of the sucker pushes it firmly against the object.

Humans have copied the design of the octopus's suckers to make objects that can stick to a smooth surface. Bathmats, soap holders, ornaments for car windows, and many other objects use rubber or plastic suction cups that resemble the suckers of an octopus.

If you have used something like this, you may have discovered that the suction cups work best when they are wet. An octopus does not need to worry about keeping its suckers wet, because they are always under water!

Taste and Smell

Instead of tasting with their tongues, octopuses have taste **receptors,** or taste-sensitive cells, all over their bodies. Octopuses' sense of taste is ten times better than ours. The most sensitive tasting areas are in the octopus's suckers. This means that when an octopus uses its suckers to touch another object, it is tasting it as well as feeling it. It does not have to bring **prey** back to its mouth to find out if it will make a tasty meal.

Octopuses do not have noses and do not have a sense of smell in the same way that humans or other land animals do. However, octopuses can taste things from a distance, similar to the way that humans can sense certain substances in the air by smelling them. When another animal is close to an octopus, a few **molecules** from it drift through the water, and the octopus detects them using its taste receptors. An octopus can sense when its next meal is within range, even while it is sitting safely in its den. This ability to sense molecules in the water is called **chemosensation.**

This giant octopus is closing in on a kelp crab. It can taste the crab before it even touches it.

How Octopuses Hear

Octopuses' sense of hearing works very differently from that of humans. Octopuses have no ears, but that does not mean that they cannot detect sounds. In fact, you might think of an octopus's whole body as a kind of ear. When sounds occur in the ocean, such as the swishing of a fish as it swims, tiny vibrations are sent out through the water. These ripples are picked up by sensitive cells all over the octopus's body. In other words, an octopus hears sounds by feeling them!

I DIDN'T KNOW THAT!

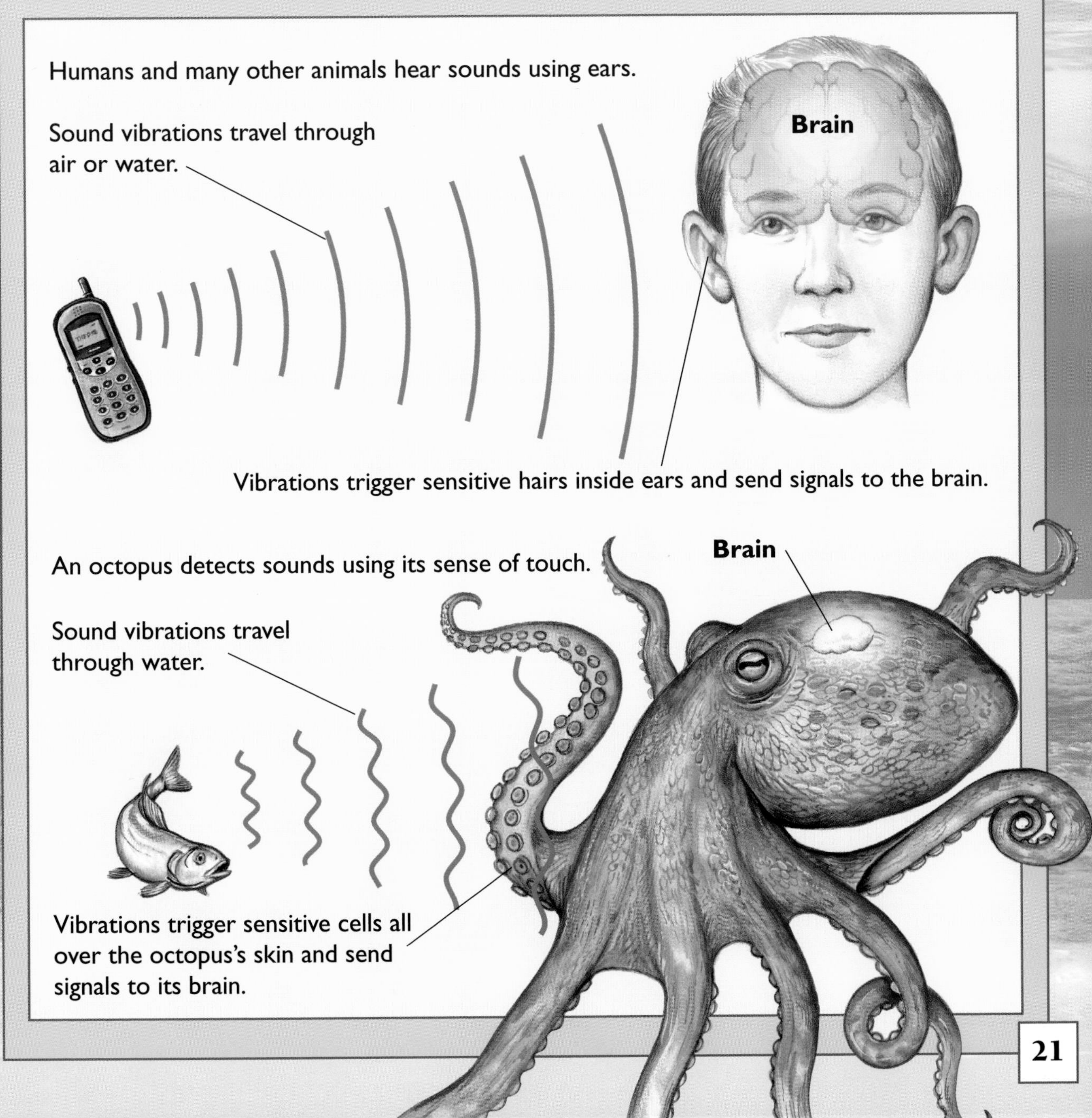

CHAPTER 4
Finding Food

Octopuses are **predators,** which means they hunt, kill, and eat other animals. They are good hunters, but they cannot move fast for long stretches at a time. So instead of chasing animals that swim quickly like fish, octopuses usually feed on slow, **benthic** creatures like crabs, lobsters, sea snails, mussels, clams, and other shellfish. Because benthic animals are often covered with a tough, protective shell, octopuses have a hard **beak** for piercing or cracking shells open. Sometimes they will also use their **suckers** to pull open mussels, clams, and scallops, which have two shell halves joined by a hinge.

In stressful situations, such as when they are being transported, octopuses sometimes eat their own legs. They may also do this when they are short of food.

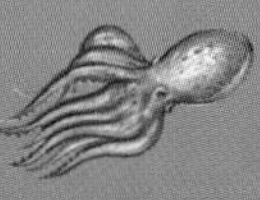

Although in some movies octopuses seem like scary monsters, in real life they do not attack people. However, it is possible for divers to drown if a very large octopus holds onto them and stops them from returning to the surface.

Because young octopuses grow very quickly, they need to eat a lot. Each day a young octopus eats about three percent of its body weight in food, and grows by about two percent.

OCTOPUS BEAK

It is usually very difficult to see an octopus's beak because it is right in the middle of its eight legs.

This common octopus is eating a crab it has caught. The crab is held firmly by the octopus's mouth, which is in the middle of its legs.

Because octopuses have such sharp senses, they do not have to try very hard to find their **prey.** An octopus can simply sit at home in its den, waiting for an animal to come close enough to reach. When it sees, feels, or tastes its prey nearby, the octopus can reach out and catch it with its suckers, then pull the food back into its mouth.

When food is scarce, however, octopuses do leave their dens to find prey. They prefer to hunt at night, or when there is not much light. They wander over the ocean floor or across a coral reef looking for food.

Octopuses' ability to change color and **camouflage** themselves makes it difficult for other animals to see them. This allows octopuses to take prey by surprise.

Catching Prey

The main way for octopuses to catch their **prey** is with their **suckers.** But some octopuses also use another method. Several **species** have thin webbing that stretches between their legs. They swim up above their prey, spread out their legs, and sink down like a parachute onto the ocean floor, trapping the prey underneath. In this way, an octopus can catch several animals at once. It then uses its suckers to pick up the food it has trapped and put it into its mouth.

▲ This common octopus has spread out its body into a parachute or umbrella shape. It will trap its prey by sinking down on top of it.

Poison Bite

To actually eat its prey, the octopus first gives it a poisonous bite. If the animal has a shell, the octopus uses its **beak** to crack it open, or it drills a hole in it with its **radula**—a kind of sharp tongue covered with tiny, hard teeth. Then it injects poisonous saliva into the animal's body.

This octopus from the Philippines is hunting along a coral reef. You can see several different types of colorful coral nearby.

The saliva contains a **toxin** that makes the prey unable to move, and digestive juices that start to break the flesh down into a gooey jelly. Once the meat has softened, the octopus can scrape it out of the shell and into its mouth using its radula. Octopuses have to eat this way because their mouths are small and they cannot chew their food.

What a Sucker!

I DIDN'T KNOW THAT!

In this photo of a lesser octopus's leg, you can see how it narrows to a thin finger shape at the tip. Some octopuses use one of their legs as bait. The octopus stays very still, then reaches out with one leg and wiggles the tip so that it looks like a small worm. This may attract a crab, fish, or other hungry sea creature. When the prey gets close enough, the octopus grabs it and pulls it into its mouth.

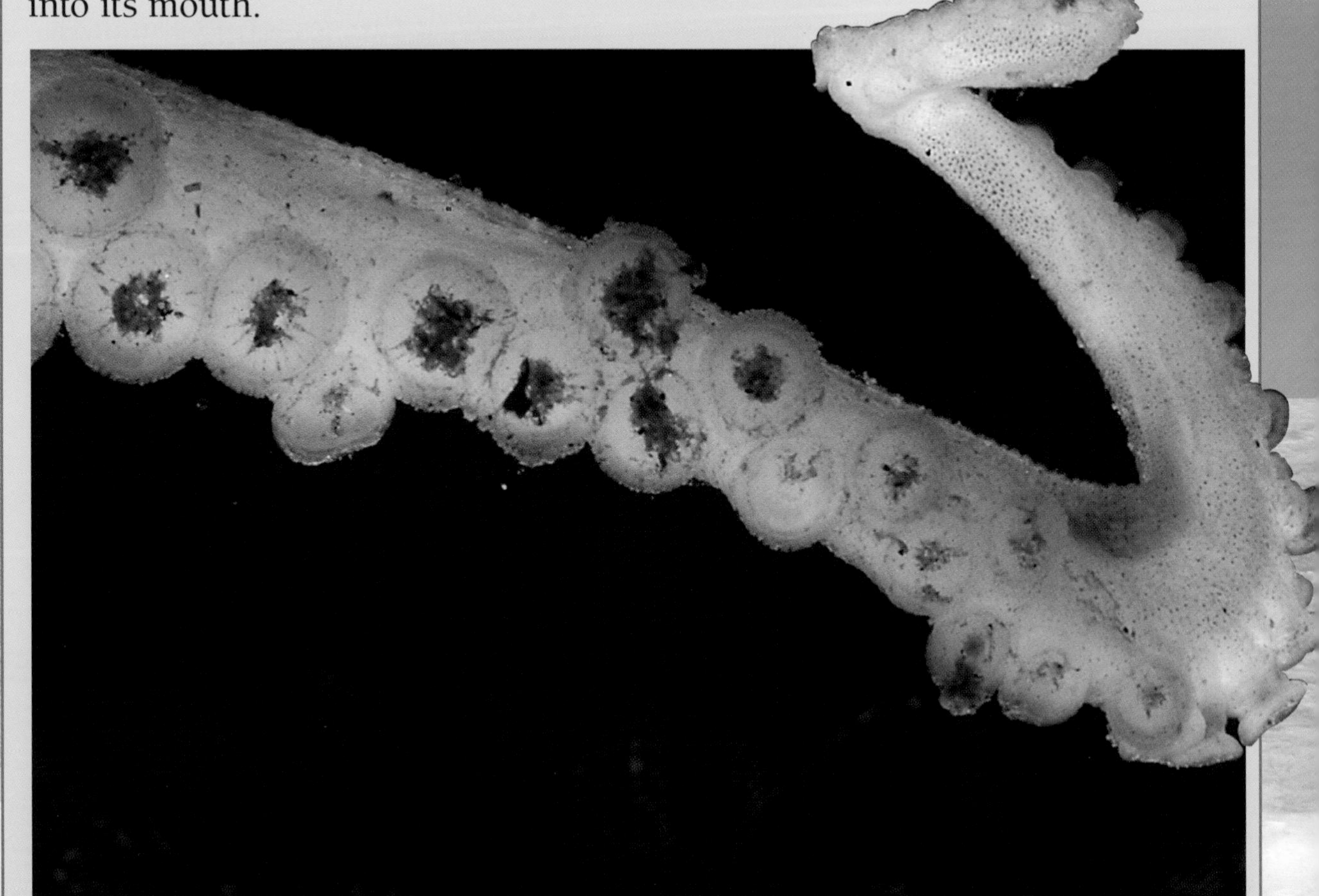

CHAPTER 5
Staying Safe

With their soft bodies and muscular legs, octopuses make an easy meal for other animals. Sharks, dolphins, seals, seabirds, and most of all, huge, fierce moray eels are constantly on the lookout for octopuses to eat. And since octopuses do not have a shell, they have to find other ways to protect themselves from being eaten.

An octopus normally tries to avoid danger entirely by hiding in its den and staying still, making it almost impossible to see. But if an octopus is frightened or threatened by a **predator**, it can sometimes make a quick escape by using its **siphon** to zoom away.

Octopuses can break off one of their legs and leave it as a decoy to distract an attacker while the octopus escapes. Even after it breaks off, the leg can still wiggle and change colors!

A previously unknown type of octopus, called the mimic octopus, was discovered off the coast of Indonesia in 1998.

The beak of the very poisonous blue-ringed octopus can pierce a diver's wetsuit.

▶ This young octopus is being swallowed by a cup coral. Corals are simple animals that stay fixed in one place and catch prey as it wanders past. This young octopus might have thought this coral was a harmless plant.

Camouflage Experts

Octopuses are also excellent at **camouflage.** Chameleons are famous for being able to change the color of their skin, but in fact, octopuses are much better at doing this than almost any other creature. Not only can an octopus change the patterns and colors on its skin to match its surroundings, it can also change the actual texture of its skin, giving it ridges or bumps to blend in with rocks and seaweed.

This midget octopus from New Zealand has changed its skin texture as well as its color in order to disguise itself as a rock covered in seaweed on the ocean floor.

Octopuses change their skin color using **chromatophores,** which are tiny bags of color just under their skin. Chromatophores come in different colors, including yellow, red, brown, and black. Octopuses can make other colors, such as orange, by using combinations of different chromatophores. Each chromatophore is surrounded by muscles that can expand it into a big spot of color. When the muscles relax, the chromatophore shrinks to a tiny dot.

The Mimic Octopus

One type of octopus, the mimic octopus, changes its appearance to imitate other animals. It will flatten itself, gather its legs together, and turn its skin brown to look like a sole (a type of fish with a very flat body). Or it may hide in its den and stick just one or two legs out, giving them black and yellow stripes so that they look very similar to sea snakes. Many of the animals the mimic octopus imitates are poisonous, so scientists think it copies them in order to scare predators away.

Defensive Weapons

Like all **cephalopods,** most octopuses make black ink inside their bodies. The ink is stored in an organ called the **ink sac.**

This octopus is squirting out ink behind it as it jets away. The octopus has also changed color to make itself difficult to spot in the cloud of ink.

When an octopus feels threatened by another animal, it uses its **siphon** to squirt ink at its attacker. It may also use a jet of water from its siphon to spread the ink out and cloud the water. This helps to confuse the **predator** while the octopus escapes. Scientists have also learned that the ink confuses predators by blocking their sense of smell for a short while.

However, the ink can also be dangerous for the octopus itself. If a captive octopus releases ink into its tank, its keepers have to change the water quickly or the octopus may die. This is not because the ink is poisonous (in several countries, people eat octopuses stewed in their own ink) but because the ink may coat the **gills,** making it difficult for the octopus to breathe. Some deep-sea octopuses do not have an ink sac. This is probably because it is very dark in deep water, so a cloud of black ink would be completely invisible.

Octopuses also use their poisonous bites to defend themselves. Most species do not produce a very strong poison. But the blue-ringed octopus from Australia has a deadly bite. Several people have died after picking up or stepping on a blue-ringed octopus, causing it to bite in self-defense.

Escape Artist

I DIDN'T KNOW THAT !

An octopus can squeeze its rubbery, squishy body through unbelievably tiny gaps. All it needs is a space big enough for its **beak** and eyeballs to get through—the rest of its body can change shape and "pour" itself through any tiny crack, tube, or opening. For example, a common octopus with a leg span of 2.5 feet (80 centimeters) and a body the size of a soccer ball could squeeze through a hole the size of a golf ball.

This is why some scientists studying octopuses often find their octopuses missing from their tanks. They have wriggled out of a tiny crack in the lid or pushed the lid aside just far enough to escape. In the photograph below, a day octopus shows its shape-changing skills by squeezing itself out of a tiny hole in the lid of a container. One way to prevent octopuses from escaping is to put a piece of carpet on the underside of the tank's lid. Octopuses' **suckers** cannot get a good grip on carpet, making it harder for them to pull themselves out.

CHAPTER 6
Reproduction

Octopuses usually live alone, so when they want to mate and have young, they first have to find a member of the opposite sex. Male and female octopuses use their color-changing abilities to send signals to each other. To show that it is ready to mate, an octopus makes beautiful waves of different colors pass across its body. Before mating, a male and female octopus will play together for a while.

The male octopus has one leg that is different from his other legs. The tip of this leg is shaped a bit like a spoon, with a groove running down the middle. The male uses this to deliver sperm cells to the female while they are mating. She needs these cells in order to get her eggs **fertilized** so they can develop.

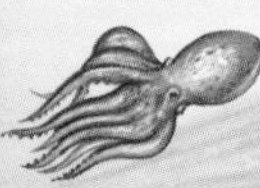

Octopus eggs are tiny compared to an adult octopus. The female common octopus is 2–3 ft (60–90 cm) across, but each of her eggs is about the size of a grain of rice.

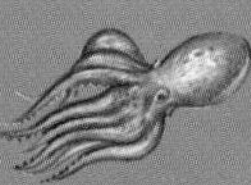

Even if an octopus lays thousands of eggs, usually only one or two will survive to adulthood.

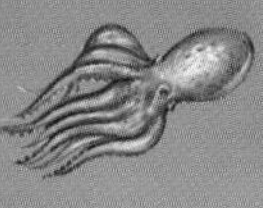

Octopus eggs are sometimes called spawn.

▶ These Maori octopuses are mating. The male (on the left) is reaching over to pass sperm cells to the female.

After mating, eggs start to grow inside the female's body. The number of eggs ranges from about 150 in the pygmy octopus, up to as many as 300,000 in the common octopus. The female lays the eggs attached to long strings of a jelly-like substance. She sticks one end of each string to the roof of her den and watches over the eggs very carefully, squirting water over them with her **siphon** to keep them clean.

A cluster of octopus eggs attached to a rock. These eggs have been exposed to the air at low tide, and the mother has left them. Usually, the eggs stay underwater and the mother watches over them at all times.

How many eggs?

Many egg-laying animals do not take care of their offspring once they are hatched. These animals usually lay a lot of eggs. That way, even if most of the eggs and young are eaten, some will still survive to become adults.

Common octopus:
Lays 200,000–
300,000 eggs

Turbot:
Lays up to
5 million eggs

Green sea turtle:
Lays about
110 eggs

Seahorse:
Lays 200 eggs

Birth and Babies

Male octopuses do not stay to look after their eggs or their young. The mother cares for them by herself. She stays in her den, guarding the eggs while they develop, which can take between four and five months. However, the eggs of giant octopuses may take as long as seven months to develop.

Self-Sacrifice

I DIDN'T KNOW THAT!

In some **species,** including the common octopus and the giant octopus, looking after the eggs takes up every second of the mother's time. She cannot go out to find food while the eggs are developing, so she slowly becomes weaker and weaker. She usually dies soon after the eggs hatch. This means that a female octopus may have only one set of offspring in her lifetime. This picture shows a female octopus guarding her eggs by wrapping her legs around them.

In this amazing close-up photo of two-spotted octopus eggs, you can see the baby octopuses inside the eggs, ready to hatch. One tiny octopus has already broken out of its egg.

Just before birth, the tiny octopuses turn around inside the eggs so that their legs are facing toward the end of the egg that is attached to the string. This means that they are ready to hatch. To help them break free, the mother prods the eggs with her legs or squirts a hard jet of water at them with her **siphon,** and hundreds or thousands of young octopuses then escape into the water.

Most newly hatched octopuses immediately go to live on the ocean floor. They find themselves a den and start hunting, just like their parents. But some species of octopus start life by floating around in the sea as **plankton.** This is the name for the millions of tiny plants and animals that drift through the ocean in huge groups. Many animals, including the biggest whales, feed on plankton, so a lot of young octopuses never make it past this stage. Those that do survive grow bigger and eventually move off to live by themselves on the ocean floor.

CHAPTER 7
Octopus Intelligence

Many experts believe that octopuses are the smartest of all **invertebrates.** Scientists also often say that octopuses seem to have their own personalities and will make eye contact with the humans who work with them. In some ways, octopuses seem to have more in common with cats or dogs than with their invertebrate relatives, such as slugs and snails.

An octopus's brain contains about 300 million neurons, or brain cells. A sea slug's brain has about 20,000 neurons, while that of a human has 100 billion.

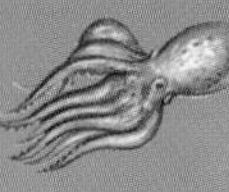

The octopus and its relatives, the squid and nautilus, are the only mollusks to have large and complex brains.

An octopus's brain surrounds the esophagus (the tube that carries food from its mouth to its stomach).

▶ Scientists have recently found that an octopus's brain power is not only in its head, but also in its legs. Each leg seems to be able to make its own decisions about what to do.

Why are octopuses so intelligent? It is partly because they have eight legs and very advanced senses. It takes a lot of brain power to deal with all of the information that an octopus collects through its senses of sight, taste, and touch, and then to send instructions to eight separate legs. So an octopus needs a very complex nervous system (the network of cells that form the brain and link it to the rest of the body).

This scientist is holding a 65-lb (30-kg) giant octopus, which was found off the western coast of Canada.

Another reason octopuses are smart might be that they do not have very much natural protection, such as a shell. Like humans, they have had to use intelligence to stay safe. Over time, this resulted in the development of bigger and more complex brains. Like a human brain, an octopus's brain is made up of several different sections that do different jobs. This is unusual for invertebrates.

Intelligence Tests

Scientists have come up with many tests to see how intelligent octopuses are and to find out what they can do. In one famous test, the **oceanographer** Jacques Cousteau gave an octopus some food in a jar with a lid. The octopus figured out how to open it to get the food. When the same octopus was given a similar food jar the next day, it opened it immediately. This showed that the octopus had remembered what it had learned.

Other experiments seem to suggest that octopuses can learn by watching others. In one test, an octopus that had never been given a food jar with a lid was allowed to watch another octopus opening one. When it was given its own jar, the first octopus opened it much more quickly than octopuses that had not been allowed to watch.

In another test, scientists trained one group of octopuses to chase balls of a particular color by letting them watch a second group of octopuses that had already been trained. In the second group, half of the octopuses had been trained

A captive common octopus peers through the glass of its aquarium.

This octopus is taking part in an intelligence test in an animal research center.

to chase red balls, and the other half had been trained to chase white balls. Some of the untrained octopuses were allowed to watch the octopuses chasing red balls, and some were allowed to watch those chasing white balls.

Finally, all of the untrained octopuses were given red and white balls to play with. It turned out that each octopus chose to chase the balls of the color it had seen octopuses chasing earlier. This way of learning by watching is called "observational learning," and it is a sign of high intelligence. However, scientists still disagree about what these experiments prove. Experiments continue to be performed to figure out how smart octopuses really are.

Another sign of their intelligence is that captive octopuses become bored quickly. They need plenty of toys and objects to keep them interested. Some species even like their keepers to pet them!

Communication and Color

Intelligent animals often have a complicated system of communication. Humans communicate using language, while chimpanzees talk to each other using calls and facial expressions. Birds and whales send messages by singing.

The bright colors of this blue-ringed octopus suggest that it might be angry or about to bite.

Octopuses use their color-changing ability to communicate. In addition to making beautiful rippling patterns to attract a mate, many octopus **species** can use their skin to signal emotions as well. For example, if a common octopus turns white, it usually means that it is scared. Red is a warning signal that shows that the octopus is angry. The blue-ringed octopus's blue rings also provide awarning—they become much brighter just before the octopus bites.

However, because octopuses do not normally live in groups, their communication system is not as complex as that of chimpanzees or other **social** animals. The only messages they send to each other are "I want a mate" or "Keep away!"

Fast and Flashy

The octopus in the photograph below has made its skin pale beige with red speckles to match the sand that it is lying on. Octopuses can change color completely in a split second and can flash through a series of different colors and patterns one after the other. One common octopus has been filmed as it changed its skin from dark greenish-brown with a rough, bumpy surface (to imitate seaweed) to smooth, bright white, all within a second. To make such a dramatic change, the octopus has to activate millions of **chromatophores** at once.

I DIDN'T KNOW THAT !

CHAPTER 8
Octopuses and People

For thousands of years, octopuses have been an important source of food for humans. They are popular in countries around the Mediterranean Sea, especially in Greece and Italy—where a whole octopus is often stewed in its own ink—as well as in East Asia. In Japan, octopus is often eaten raw, and in Korea, people like to eat octopuses that are so fresh that their legs are still wiggling.

Octopus was an everyday food for the ancient Greeks and ancient Romans. They stewed octopuses in olive oil and herbs.

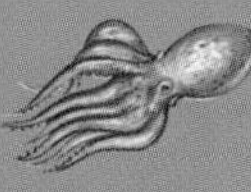

Traditional octopus-catching pots were made of clay. Today they are usually made of plastic.

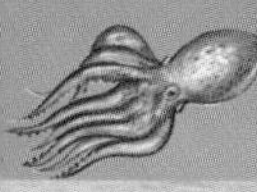

Fishing boats sometimes catch the remains of giant sea creatures resembling squid or octopuses. Based on these remains, some people think that there may be giant octopuses with a leg span of up to 200 ft (60 m).

▶ In Greece octopuses are hung up to dry in the sun before they are barbecued.

A market stall in Japan displays a large selection of freshly caught octopus.

Throughout history, sailors have often reported seeing huge, terrifying creatures at sea, with the strength to drag a ship to the ocean floor. Such stories about sea monsters could have started after someone was frightened by a larger than usual octopus.

People eat many different species of octopus, but the common octopus is one of the most popular kinds. Each year, humans catch about 60,000 tons of common octopuses—that's more than 30 million of them. Octopus has a rubbery texture and tastes slightly salty and fishy.

Octopuses were mentioned by the ancient Greek philosopher, Aristotle, over 2,000 years ago. They also are part of the myths and legends of many cultures. Ancient peoples of the Mediterranean believed in a sea god with many legs called Yamm. In Tahiti, there is a legend of a sea-demon called Rogotumu who drags people to the ocean floor with his legs. In the folklore of the Bahamas, the Lusca is a terrifying eight-legged sea creature that can change its shape and color.

This old drawing shows an enormous octopus-like monster attacking a sailing ship. In real life it is extremely unlikely that an octopus could grow this big, but the picture does show that sailors used to be very afraid of octopuses.

Keeping Octopuses

Octopuses are difficult to keep in captivity. They need to live in tanks of salt water with plenty of space and hiding places. They also need to be fed the right food (such as live crabs), or else they may not survive. But some octopuses do live happily in zoos, aquariums, and science labs. Special suppliers raise octopuses for scientists to study, and the suppliers provide the right food for the octopuses, too. Scientists often study octopuses because they help us to learn more about animal behavior and intelligence.

Some people want to keep octopuses as pets. Feeding and caring for a pet octopus can be very expensive. Keeping octopuses can also be dangerous. People who think it would be exciting to own an exotic species, such as a blue-ringed octopus, are taking a big risk, especially if they try to annoy their octopus into changing colors or try to lift it out of the tank. The blue-ringed octopus bites easily and its poison can kill a human within two or three minutes.

Getting so close to a blue-ringed octopus is risky—they are poisonous. Most experts say that divers should avoid them at all costs.

The easiest octopuses to keep in captivity include the common octopus and the pygmy octopus, which are calmer species. They are not likely to attack a human and do not have such a dangerous bite.

Octopuses in Danger?

No octopus species are officially listed as endangered. However, it is very difficult to collect information about octopuses because they are so hard to see and count. With so many octopuses being caught by humans for food, it is possible that some **species** are endangered or have even become extinct.

Populations of some species, such as the mimic octopus (below) and the blue-ringed octopus, might be endangered in particular areas where people collect them to be sold as pets. Experts recommend that pet octopuses should never be taken from the wild, but should only be bred in captivity.

Glossary

beak hard structure surrounding an octopus's mouth

benthic living on or occurring on the ocean floor

camouflage color, shape, or pattern that disguises an animal by causing it to blend in with its background

cephalopod type of mollusk that has eight or more arms with suckers surrounding a mouth with a beak

chemosensation ability of octopuses, or other water-dwelling animals, to gain information about their surroundings by detecting molecules floating in the water

chromatophore tiny sac or bag of color under an octopus's skin

ectothermic having a body temperature that warms up or cools down along with the outside temperature; cold-blooded

fertilized enabled an egg to develop by joining it with a male sex cell

focus sharpen an image so it is clear

gill breathing organ that absorbs oxygen from the water. Many underwater creatures have them, including octopuses, crabs, and fish.

ink sac organ where an octopus stores its ink

invertebrate animal with no backbone

mantle thick, leathery layer that covers and protects an octopus's organs

mantle cavity chamber inside the visceral mass of an octopus that contains the gills

midden pile of shells and other remains found outside an octopus's den

molecule grouping of two or more joined atoms. An atom is the smallest unit of a substance that has all of the properties of that substance.

mollusk type of invertebrate with a soft body, a toothed tongue called a radula, and usually a protective shell. Although octopuses and slugs are mollusks, they have no shell.

nocturnal active mainly at night

oceanographer scientist who studies the oceans

plankton tiny sea creatures, including both plants and animals, that float together in large groups throughout the ocean

predator animal that hunts other animals for food

prey animal that is hunted by another animal for food

radula tongue covered in tiny teeth

receptor sensitive cell that detects certain kinds of information in an organism's surroundings, such as sounds, tastes, or sensations

reproduction process of mating and producing offspring

retina patch of light-sensitive cells found at the back of the eyeball

siphon tube extending from an octopus's body that takes in breathing water and lets out waste water. It may also be used to squirt ink or to propel the octopus through the water very quickly.

social tending to live in groups

species group of organisms that share certain features and that can breed together to produce offspring that can also breed

suckers round, cup-shaped disks on an octopus's legs, used for feeling, holding, and even tasting objects

tentacle one of two extralong legs of squids and cuttlefish. Sometimes this word is used to refer to one of an octopus's legs, but this is not really accurate.

toxin another word for a poison

vacuum space with nothing at all inside, not even air

visceral mass part of an octopus, or any cephalopod, that extends from behind the head, contains the internal organs, and is covered by the mantle

Further Reading

Hirschi, Ron. *Octopuses*. Minneapolis: Carolrhoda Books, 2000.

Hunt, James C. *Octopus and Squid*. Monterey, Calif.: Monterey Bay Aquarium Foundation, 1997.

Stille, Darlene R. *Octopuses*. Chicago: Heinemann Library, 2003.

Swanson, Diane. *Welcome to the World of Octopus*. Portland, Ore.: Graphic Arts Center Publishing Co., 2000.

Trueit, Trudi. *Octopuses, Squids and Cuttlefish*. Danbury, Conn.: Franklin Watts, 2002.

Index

Numbers in *italic* indicate pictures